COCOA BEANS

by Amy C. Rea

Cody Koala

An Imprint of Pop!
popbooksonline.com

abdobooks.com
Published by Pop!, a division of ABDO, PO Box 398166, Minneapolis,
Minnesota 55439. Copyright © 2022 by Abdo Consulting Group, Inc.
International copyrights reserved in all countries. No part of this book may
be reproduced in any form without written permission from the publisher.
Cody Koala™ is a trademark and logo of Pop.

Printed in the United States of America, North Mankato, Minnesota.

052021
092021

THIS BOOK CONTAINS
RECYCLED MATERIALS

Cover Photos: Shutterstock Images, foreground, background
Interior Photos: Shutterstock Images, 1 (foreground), 1 (background), 5,
9, 15, 17 (bottom left), 17 (bottom right), 19, 21; Red Line Editorial, 7 (top);
iStockphoto, 7 (bottom left), 7 (bottom right); Simon Rawles/Alamy, 11; Jake
Lyell/Alamy, 12; Mile 91/Ben Langdon/Alamy, 17 (top); Mariusz Szczawinski/
Alamy, 18

Editor: Aubrey Zalewski
Series Designers: Laura Graphenteen and Colleen McLaren
Library of Congress Control Number: 2020948879
Publisher's Cataloging-in-Publication Data
Names: Rea, Amy C., author.
Title: Cocoa beans / by Amy C. Rea
Description: Minneapolis, Minnesota : Pop!, 2022 | Series: How Foods Grow |
    Includes online resources and index.
Identifiers: ISBN 9781532169793 (lib. bdg.) | ISBN 9781098240721 (ebook)
Subjects: LCSH: Cacao beans--Juvenile literature. | Cocoa tree--Juvenile
    literature. | Chocolate industry--Juvenile literature. | Agriculture--
    Juvenile literature. | Food crops--Juvenile literature.
Classification: DDC 631.5--dc23

Hello! My name is

Cody Koala

Pop open this book and you'll find QR codes like this one, loaded with information, so you can learn even more!

Scan this code* and others like it while you read, or visit the website below to make this book pop.

popbooksonline.com/cocoa-beans

*Scanning QR codes requires a web-enabled smart device with a QR code reader app and a camera.

Table of Contents

What Are Cocoa Beans?

Cocoa beans are seeds.

They come from cacao trees.

Cocoa beans grow in large

pods. Each pod has 20 to

60 seeds.

Cacao trees grow a fruit that can be eaten. The seeds are inside the fruit.

Watch a video here!

Growing Cocoa Beans

Cocoa beans grow in countries near the **equator**. The equator runs through parts of South America, Africa, and Southeast Asia.

Where Cocoa Beans Grow

1. Mexico
2. Dominican Republic
3. Colombia
4. Ecuador
5. Peru
6. Brazil
7. Sierra Leone
8. Côte d'Ivoire
9. Ghana
10. Togo
11. Nigeria
12. Cameroon
13. Indonesia
14. Papua New Guinea

Learn more here!

Cocoa beans need warm, wet weather. They grow in rain forests. The trees grow their first pods when they are approximately five years old.

Nearly half of the world's cocoa grows in the African country Côte d'Ivoire.

Chapter 3

Harvesting the Beans

Cocoa pods grow from the trunks and branches of cacao trees. The pods ripen after five to six months. Workers **harvest** ripe pods in the fall and spring.

Learn more here!

Workers cut the pods off the trees. Some pods are high up. Workers pull them down with poles. The poles have long hooks on the end.

The beans stay in the pods for approximately one week after they are picked. Then workers split the pods in half. They use clubs or knives. Workers take the beans out of the pods.

Chapter 4

Preparing the Beans

Workers cover cocoa beans with banana leaves. **Pulp** around the beans **ferments** for one week. This makes the beans less **bitter**. Then they dry in the sun.

Complete an activity here!

Machines **roast** the cocoa beans. A different machine then removes the beans' shells.

The beans are ground into a paste. People use this to make chocolate foods.

The inside of a cocoa bean is called a nib.

Some paste is pressed and dried even more. This makes a thick block of cocoa. Machines grind it into cocoa powder. Cocoa powder is an **ingredient** in many popular baked foods.

Making Connections

Text-to-Self

Have you ever eaten a food with chocolate in it?
If so, did you like it? If not, would you like to try
chocolate?

Text-to-Text

Have you read books about how other foods grow?
How were those foods different from cocoa beans?

Text-to-World

Cocoa beans grow best in warm, wet places.
What other foods grow well in these places?

Glossary

bitter – not sweet.

equator – an imaginary line that runs around the middle of Earth.

ferment – to go through a process in which the sugars in food change into alcohol.

harvest – to gather or pick crops.

ingredient – one substance used in a mixture.

pulp – the soft and juicy part of a fruit.

roast – to cook in a dry heat in an oven or over a fire.

Index

Online Resources

popbooksonline.com

Thanks for reading this Cody Koala book!

Scan this code* and others like it in this book, or visit the website below to make this book pop!

popbooksonline.com/cocoa-beans

*Scanning QR codes requires a web-enabled smart device with a QR code reader app and a camera.